Measure by Measure

by Marilyn Deen

Consultant:
Adria F. Klein, PhD
California State University, San Bernardino

CAPSTONE PRESS
a capstone imprint

Wonder Readers are published by Capstone Press,
1710 Roe Crest Drive, North Mankato, Minnesota 56003.
www.capstonepub.com

Copyright © 2013 by Capstone Press, a Capstone imprint. All rights reserved.
No part of this publication may be reproduced in whole or in part, or stored in a retrieval system, or transmitted in any form or by any means, electronic, mechanical, photocopying, recording, or otherwise, without written permission of the publisher. For information regarding permission, write to Capstone Press, 1710 Roe Crest Drive, North Mankato, Minnesota 56003.

Library of Congress Cataloging-in-Publication Data
Deen, Marilyn.
  Measure by measure / Marilyn Deen. — 1st ed.
    p. cm. — (Wonder readers)
  Includes index.
  ISBN 978-1-4296-9607-4 (library binding)
  ISBN 978-1-4296-7930-5 (paperback)
  1. Measurement--Juvenile literature. I. Title.
  QA465.D44 2013
  530.8—dc23   2011022022

Summary: Simple text and color photographs illustrate the difference between standard and metric measurements.

Note to Parents and Teachers

The Wonder Readers: Mathematics series supports national mathematics standards. These titles use text structures that support early readers, specifically with a close photo/text match and glossary. Each book is perfectly leveled to support the reader at the right reading level, and the topics are of high interest. Early readers will gain success when they are presented with a book that is of interest to them and is written at the appropriate level.

Printed in the United States of America in North Mankato, Minnesota.
042012   006682CGF12

Table of Contents

Measuring 4
Length 6
Distance 10
Weight 12
Temperature 14
Measuring Glossary 18
Now Try This! 19
Internet Sites 19
Index 20

Measuring

This boy is using his feet for measuring. This shape is about 20 feet around. But the measurement would change for someone who had larger or smaller feet.

That is why people came up with systems that would always give the same measurement no matter who did the measuring. Today the standard and metric systems are most commonly used for measuring. Most scales show both measurements.

Length

This girl is 4 feet and ½ inch tall. Feet and inches measure how long, tall, or wide something is. Feet and inches are part of the standard measurement system.

Centimeters measure the length, height, or width of something. Centimeters are part of the metric measurement system. A centimeter is smaller than an inch.

In standard measurement, 36 inches equal 1 yard. Football fields and fabric are measured in yards. An adult's arm is usually about 1 yard from shoulder to fingertips.

In metric measurement, 100 centimeters equal 1 meter. A meter is just slightly longer than a yard. Races at swimming and track meets are usually measured in meters.

Distance

Whether people travel on foot, by car, or in a plane, they can measure how far they go. In the United States, long distances are measured in miles. That is the standard measurement.

In almost every other country in the world, the metric system is used. Long distances in these countries are measured in kilometers. A kilometer is the same as 1,000 meters.

Weight

The standard system uses ounces, pounds, and tons for measuring weight. This boy weighs 50 pounds.

The metric system uses grams, kilograms, and metric tons for measuring weight. A gram is less than 1 ounce. A kilogram is about the same as 2 pounds. A metric ton is about the same as a standard ton.

Temperature

We measure temperature by counting in degrees. The hotter it is, the more degrees there are. The standard system measures degrees on the Fahrenheit scale. Water freezes at 32 degrees Fahrenheit.

The metric system measures temperature using degrees on the Celsius scale. Measuring this way, the temperature at which water freezes is 0 degrees Celsius.

There are many ways to measure this car. Measurements tell how long it is, how fast it goes, or how far it can travel.

The world is full of interesting things to measure—and different ways to measure them. You can learn a lot that way. Just pick a measurement system, and get started!

Measuring Glossary

Standard Measurement

12 inches = 1 foot

36 inches = 3 feet = 1 yard

63,360 inches = 5,280 feet = 1,760 yards = 1 mile

16 ounces = 1 pound

2,000 pounds = 1 ton

water freezes = 32 degrees Fahrenheit

water boils = 212 degrees Fahrenheit

Metric Measurement

100 centimeters = 1 meter

100,000 centimeters = 1,000 meters = 1 kilometer

1,000 grams = 1 kilogram

1,000 kilograms = 1 metric ton

water freezes = 0 degrees Celsius

water boils = 100 degrees Celsius

Now Try This!

Guess the length of your pencil in both inches and centimeters. Then measure the pencil using both standard and metric rulers. How close did your guesses come to the actual measurements? What other measurement tools are in the room or the building? Practice measuring with them.

Internet Sites

FactHound offers a safe, fun way to find Internet sites related to this book. All of the sites on FactHound have been researched by our staff.

Here's all you do:

Visit *www.facthound.com*

Type in this code: 9781429696074

Check out projects, games and lots more at
www.capstonekids.com

Index

distance, 10–11, 16
   kilometers, 11
   miles, 10

length, 6–9, 16
   centimeters, 7, 9
   feet, 6
   inches, 6, 7, 8
   meters, 9, 11
   yards, 8, 9

scales, 5

temperature, 14–15
   Celsius, 15
   Fahrenheit, 14

weight, 12–13
   grams, 13
   kilograms, 13
   metric tons, 13
   ounces, 12, 13
   pounds, 12, 13
   tons, 12, 13

Editorial Credits
Maryellen Gregoire, project director; Mary Lindeen, consulting editor; Gene Bentdahl, designer; Sarah Schuette, editor; Wanda Winch, media researcher; Eric Manske, production specialist

Photo Credits
Photos by Capstone Studio: Karon Dubke, except Shutterstock/Alfonso de Tomas, 11; Danny E. Hooks, 8; Sven Hoppe, 7

Word Count: **403** Guided Reading Level: **K** Early Intervention Level: **19**